MATH MIND BENDERS®

ARITHMETIC: BOOK B-1

DEDUCTIVE REASONING IN MATHEMATICS

by
ANITA HARNADEK

MIDWEST PUBLICATIONS
P.O. Box 448
Pacific Grove, CA 93950-0448
ISBN 0-89455-376-3

Teaching Suggestions, Answers, and Detailed Solutions

About This Series

Math Mind Benders® are worked like crossword puzzles, but each square is filled with a digit rather than a letter.

No puzzle has routine clues such as "13 × 96." Instead, the clues are interrelated with each other and a given story. Clear thinking and deductive reasoning must be used to find potential answers and eliminate those that fail to meet all conditions. Although several choices may at first seem equally acceptable, each puzzle has a **unique** solution.

Difficulty level cannot be determined from the size of the grid or the number of clues, for some puzzles in all booklets use the same 4 x 4 grid and the same number of clues.

These puzzles present a wealth of opportunity for discovering interesting facts about numbers and their relations to each other, at times merging with some basic principles of elementary number theory.

The use of calculators should be encouraged at all levels.

Each booklet includes a complete set of answers—explanations as well as grids.

Current titles in the **Math Mind Benders®** series are

Arithmetic:

- **Warm Up** (introductory)
- **Book A-1** (easy)
- **Book B-1** (medium)
- **Book C-1** (hard)

Reproduction rights for each booklet are limited to a single classroom.

About This Booklet

The reasoning involved, as well as the number and obscurity of interrelated clues, is more complex here than in the **A** booklet.

It is strongly suggested that students master the materials in both the **Warm Up** and **A** booklets of **Math Mind Benders®** before being exposed to these problems.

Teaching Suggestions

• Start your students off with the **Warm Up** booklet. Make sure that they have mastered this material before moving them to a higher level.

• Avoid adding to the inherent difficulties of these puzzles, and encourage your students to try more puzzles and thus increase their abilities to reason out answers to new challenges. **Advocate the use of calculators.**

• Conclusions that are obvious to you may not be obvious to your students. Be patient. It encourages your students to be lazy about developing their reasoning powers when you always explain how to go about finding an answer. Instead, ask logical questions designed to help your students find the reasoning process themselves.

• Having students explain their reasoning is different from having you explain yours. Your students *expect* you to be able to figure things out (you're a *teacher*!), but they're more likely to try to think for themselves when they know how other students solved the problems.

• You may find that you would like to point to a certain part of a problem and have the whole class see it (for example, pages 9–11 of **Warm Up**). Make an overhead transparency of a page you think might be especially troublesome to your students.

• Some students enjoy the challenge of working alone on such puzzles, while others would rather be part of a team effort. Consider allowing both kinds of work simultaneously.

Answers

1.

1	3	
5		1
	9	3

2.

1	1		6	4	9
2	4		6	7	9
	6	8			
2	4	5		3	7
1	1	0		8	2

3.

1	2		4	8
1	8		3	6
	8	6	4	
2		1	8	9
3	7	8		2

4.

1	1		5	8	
6	7	1		1	6
	4	5			4
3			6	3	
3	1		1	7	6
7	2	9		2	0

5.

	5	1		7	1	1
1	0	0		3	9	0
3	1	5			4	8
6	8		3	7		
		9	3	6	1	
9	9	9		9	4	1
3	1		2	5	0	9

6.

3	1		1	7
4	8		4	4
	7	5	9	
1		7	6	8
1	3	6		8

7.

1	6	
9		7
	1	2

8.

4		2	5
4	1		7
4	3	8	
4		7	3

9.

1	4		1	6
1	8		1	7
	3	1	5	
5		3	9	1
2	7	0		9

Answers [continued]

10.

7	2		2	8	3	5
1	5	2		2	2	2
	2	2	1		8	4
4		3	3	6		
9	6		5	4	7	2
7	5		3	0	6	4
	6	7	6	5		6

11.

2	3		5	4	
1	4	8		9	8
	8	4			7
5			6	3	
3	3		1	4	8
2	7	4		4	3

12.

1	8		2	1
6	3		3	0
	1	3	3	
8		1	1	1
2	3	1		9

13.

6		7	9
3	3		8
5	6	1	
4		8	4

14.

1		5	2
3	9		6
5	1	3	
2		6	7

15.

6	2		6	3	1	3
8	0	4		2	1	4
	3	3	2		7	2
1		8	3	6		
6	5		7	6	5	1
7	5		5	3	9	3
	2	3	1	4		5

16.

1	3		9	1	
6	4	8		5	4
	5	4			7
9			7	1	
3	1		5	9	4
1	0	5		5	8

Detailed Solutions

1. From 1-A, 3-A, 4-A, and 5-D, we see that three of the ages are one-digit numbers, and the other age is a two-digit number (but not 10, since 2-D can't be 0). The only possibilities are 1, 5, 9, 13; or 2, 5, 8, 11; or 3, 6, 9, 12; or 5, 7, 9, 11. 1-D rules out the second possibility, since 3 is not a factor or 2, 5, or 8. If the numbers are 3, 6, 9, 12, then (because 1-A is 12) 1-D is 13, 16, or 19, none of which is possible.

   In both other cases (1, 5, 9, 13; or 5, 7, 9, 11), 1-D implies that 5-D is 9, 3-A is 5, and 1-D is 15, forcing 4-A to be 1 or 7 and 1-A to be 13 or 11. Then 4-D is 13 or 77, making 5-A either 93 or 97. Then 5-A implies that Great-aunt Martha is either 78 or 158. Her age is less than 126, so she is 78. Working backwards from this, answers are forced for 5-A, 4-D, 4-A, 1-A, and 2-D.

2. Combine clues 1-A and 5-D, along with the fact that no answer begins with 0. (So 2-D can't begin with 0.) Then 1-A is 11, and this forces the answers to 5-D, 3-A, 13-A, and 3-D.

   From 7-A, the square of some number is in the 500s. Only two numbers, 23 and 24, qualify. Adding 103 to the squares (529 and 576), we get 632 and 679. Only 679 fits in 7-A. This forces the answers to 6-A (from 7-A), 4-D, 11-D, 1-D, 12-D, 11-A, 8-A, 14-A, 9-D, 10-A, 2-D, and 10-D.

3. 12-A implies that only 7 works for 13-D. This forces 12-A, 10-A, 11-D (because the last digit of 9-D is 3, so the last digit of 11-D is 2), 9-D, 14-A, 9-A, and 3-D.

   6-A and 3-A must be 33 and 44, or 36 and 48. 5-A implies that 6-A is even so answers are forced for 6-A, 3-A, 5-A, 4-D, 7-A, 8-D, 2-D, 1-D, and 1-A.

4. Only 3 and 4 have a two-digit cube, so answers are forced for 10-A, 8-D, 7-A, 12-D, 4-D (the only two-digit square ending in 1 is 81), 19-D, 3-D, and 3-A.

   Because of 18-A, 11-A is not 1. A calculator quickly shows that the sixth power of 2 has two digits while the sixth power of 4 has four digits. Thus answers are forced for 11-A, 18-A, 5-A, 1-A, 1-D, 11-D, 2-D, 9-A and 6-D (only one number works here), 15-D (because Angelo caught fewer than 6-D), 14-A, 17-D, 13-D, 20-A, 12-A, and 16-A.

5. From 6-A and 13-D, we see that 2-D must be 105. This forces 6-A, 13-D, and 8-A.

   6-D is in the 130s, so 10-A is in the 60s. This gives us both 6-D and 10-A, as well as forcing 1-A, 1-D, 9-A, 20-A, 20-D, and the first digit of 16-D.

   15-A has to be 999, since it is the only 3-digit number ending in 99 that is divisible by 27. This forces answers to 11-A, 19-A, 15-D, 16-D, 18-D, 12-D, 13-A, 11-D, 5-D, 3-D, 14-D, 17-A, 7-A, 4-D, and 3-A.

6. From 2-D, 12-A, and 11-D, the factors of 3-D include 8, 11, and 17. The product of these factors is 1,496, too close to 1,500 to allow any additional prime factors. (For example, 1,496 × 2 would be too many pieces for a "1500 pieces" puzzle.) Thus, answers are forced for 3-D, 2-D, 11-D, 14-A, 12-A, 13-D, 9-D (because, from first digit of 3-A, Olivia is not yet 20, and Luis is younger than she), and 9-A.

   Half of 3-D = 748, so 10-A is at least this much. Adding 13-D and 3-A to 748 will not put 10-A in the 800s, so 10-A is 768, and 3-A is 17.

   The length of the puzzle (6-A) is in the 40s. Breaking 1,496 (3-D) into prime factors of 2, 2, 2, 11, and 17, the only possibility for 6-A is 44, forcing 1-D to be 34, and forcing 1-A, 5-A, and 4-D.

7-A, whose first and last digits are known, has a factor of 2 (i.e., 1-A − 14-A), and only 33 × 23 = 759 gives the required result. This also forces 8-D.

7. Combining 3-A, 4-A, 5-A, and 2-D, we see that three of the ages are one-digit numbers, and the fourth age is a two-digit number. Using the given differences of 1, 2, and 3 years, the only possibilities for the four ages are 4, 5, 7, 10; or 5, 6, 8, 11; or 6, 7, 9, 12. 4-D implies that 2-D is even.

   Suppose the ages were 4, 5, 7, and 10. Then 2-D would be 4, 5-A would be 10, and 1-A would be 14, 24, ···, or 94, half of 1-A would be 7, 12, 17, ···, 47; 3-A would be 5 or 7, and 4-A would be 7 or 5. In none of these cases would 4-D work out correctly. Therefore, the supposition has to be wrong, and so the ages are not 4, 5, 7, and 10. Similarly, the ages are not 5, 6, 8, and 11. (Try first with 2-D = 6, then with 2-D = 8.)

   So the ages are 6, 7, 9, and 12, with 2-D = 6, 5-A = 12. Then 3-A is 7 or 9, 4-A is 9 or 7. But 92 (for 4-D) does not have a factor of 7, so 3-A is not 7. So 3-A is 9, forcing answers for 4-A, 5-D, 4-D, 1-A, and 1-D.

8. The average cost of the books was $1.96/4, or 49¢. The prices were equally spaced, so two of the books cost more than 49¢. There are three equal price gaps between the costs of the books, so $1\frac{1}{2}$ of these gaps would be above the average price. It is given that one book cost 57¢, or 8¢ more than the average of 49¢. If this were the most expensive book, then the 8¢ would represent $1\frac{1}{2}$ gaps, a contradiction, since $8 \div 1\frac{1}{2}$ is not a whole number. Thus, 57¢ must be next to the highest price, and the 8¢ represents 1/2 of the price gap between books. Thus, the gap is 16¢, and the prices are 73¢ (i.e., 57¢ + 16¢), 57¢, 41¢, and 25¢. So 10-A is answered, also answering 11-D.

   Only two of the other costs fit in 2-A and 3-D, so these are forced, as are 2-D, 6-A, 4-A, 5-D, 8-D, 9-A, 1-D, 1-A, and 7-A.

9. 5-A cannot begin with 0, so 1-D is not 10. But 1-D cannot exceed 11, for Beaufort, who is 8 years older, is still a teenager. Then answers are forced for 1-D, 11-D, 14-A, 6-A, 3-A, 4-D, 3-D, 9-D, 9-A, 8-D, 10-A, 12-A, 13-D, and 7-A.

   See 2-D. The sum of 3-A and 13-D is 23. Since no factor of 12 divides 23, 23 must be a factor of 2-D, and 12 must be a factor of 1-A × 5-A. 2-D ends in 3 and has a factor of 23, so the second digit of $\frac{\text{1-A} \times \text{5-A}}{12}$ must be 1. Then the only possibilities are 11 × 23 = 253, 21 × 23 = 483, 31 × 23 = 713, and 41 × 23 = 943. These products would force 1-A and 5-A, respectively, to be 12 and 15 (no, since 12 divided into 12 × 15 is not 11), or 14 and 18 (yes), or 17 and 11 (no, since 12 does not divide 17 × 11), or 19 and 14 (no, since 12 does not divide 19 × 14).

10. The lowest score is in the range 51–75 (i.e., 99 − 24). Three students who did not have the lowest score are named, leaving Cora with the lowest score. Cora's score is divisible by 12 (from 25-A), so 1-A must be 72 (not 60, since 2-D cannot start with 0), and 25-A is 6.

    The given information then implies that the four scores are 72, 76, 84, and 96. This forces 5-D, 12-A, 18-A, 22-A, 19-D, 20-D, 4-D, 16-A (the other three scores have been placed), 11-D, 7-A, 1-D, 13-D, 13-A, 17-D, 21-A, and 9-A.

    23-A is in the range 6,160–6,969 and has a factor of 615 (i.e., 5 × 123). Since 6,160 ÷ 615 > 10, and 6,969 ÷ 615 < 12, the missing factor is 11. So 23-A = 5 × 123 × 11 = 6,765, and this forces 24-D, 2-D, 10-A, 8-D (from definition of 14-A), 14-A, 15-D, 3-A, 3-D, and 6-D.

11. 5-A, 11-D, 11-A, 14-A, and 15-D are answerable immediately. The only costs that fit 3-A, 4-D, 7-A, and 8-D are, respectively, 54, 49, 98, and 87. This forces 3-D, 10-A, 6-D, 12-D, and 12-A.

    For 1-D to end in 1, 19-D has to be 4, forcing answers for 1-D and 18-A. 1-A is 21–29, and 9-A is 14–94, so 1-A + 9-A is in the range 35–123. Adding 241, the range of 2-D becomes 276–364. Since the second digit of 2-D is 4, the first digit has to be 3, so 1-A = 23, and only final digits of 1-A, 9-A, and 241 need be used to determine that 2-D = 348 and 9-A = 84. This forces 13-D to be 344.

    Since 20-A and 17-D share their last digit, their sum is even. Adding 22 to this sum, 16-A is even. Only 16-A = 148 will make 17-D enough so that 17-D + 20-A + 22 = 16-A. Then 17-D + 20-A = 148 − 22 = 126, so 17-D and 20-A have to be 83 and 43.

12. Start with 14-A and 4-D. This also forces 11-D. From 13-D, 21 is the only number for 3-A, forcing 3 for 13-D.

    3-D implies that the first digit of 10-A is 1. The last digit of both 10-A and 3-A is 1, so 3-D ends in 1, and 10-A is 111. Then answers are forced for 3-D, 6-A, 7-A, 1-A, 12-A, 2-D, 8-D, 5-A, 1-D, 9-D, and 9-A.

13. There are seven possibilities for 2-A and 3-D: 13 and 32, 24 and

[Continued on page 28]

In the clues, "A" means across and "D" means down. For example, "4-D" would refer to clue number 4 DOWN.

Each square takes a single digit from 0 through 9. No answer begins with 0.

1. Not necessarily listed in the order of their ages, Caleb, Donna, Eduardo, and Lauren are four children Rita baby-sat last week. The differences between successive ages of the children were all the same.

Rita's Great-aunt Martha, who lives with Rita and is not older than 125, offered to take Rita's place baby-sitting the youngster who had the measles, but Rita turned down the offer, knowing that Great-aunt Martha would rather go to a duplicate bridge game that evening.

ACROSS

1. Donna's age
3. Caleb's age
4. Lauren's age
5. Half of age of Great-aunt Martha, with digits reversed

DOWN

1. 3-A × 1/3 of 5-D
2. Times Rita baby-sat two weeks ago
4. 1-A × 4-A
5. Eduardo's age

1	2	(shaded)
3	(shaded)	4
(shaded)	5	

In the clues, "A" means across and "D" means down. For example, "4-D" would refer to clue number 4 DOWN.

Each square takes a single digit from 0 through 9. No answer begins with 0.

2. Polly lives with her mother and father, her three brothers, Khalid, Joe, and Marcos, her two sisters, Susan and Mitra, the family's tiger-striped cat, Tiger, and the family's dog, a Dalmation named Spot.

Although Tiger and Spot get along well themselves, they don't usually like each other's friends very much. Tiger thinks Spot's friends are a noisy bunch who enjoy roughhousing and getting dirty, while Spot thinks Tiger's friends are too lazy to do much other than walk dantily or sit around washing their faces or just lie napping in the sun.

ACROSS

1. Polly's age
3. 1-A × 59
6. 2 less than number of students in Polly's class at school
7. 103 more than 6-A × 6-A
8. Spots on Spot
10. 1-D + 3-D + 5-D + 8-A
11. Stripes on Tiger $\left(\frac{8\text{-A} + 6}{2}\right)$
13. 1-A × 10
14. Age of father of Polly's Great-aunt Martha

DOWN

1. Age of Polly's mother 25 years ago
2. Polly's favorite number
3. Sum of 1 + 2 + ⋯ + Polly's age
4. Age of Polly's father
5. 9 × 1-A
9. 10 × sum of 4-D and 11-D
10. Joe's age
11. Age of Polly's father 9 years ago
12. 6-A × 3

Fill in the clues from the previous page on the puzzle grid below.

2.

1	2		3	4	5
6			7		
	8	9			
10				11	12
13				14	

In the clues, "A" means across and "D" means down. For example, "4-D" would refer to clue number 4 DOWN.

Each square takes a single digit from 0 through 9. No answer begins with 0.

3. Bob eats with a healthy appetite, but he's such an active person that he's still on the thin side. His father teases him about having hollow legs, but his mother says she thinks his arms are hollow, too.

Bob's Great-aunt Martha offered to take midmorning and mid-afternoon snacks to him at his school, but he regretfully said no, because he isn't allowed to eat during classes or in the hall between classes.

ACROSS

1. Bob's age four years ago
3. See 6-A
5. Half of 6-A
6. Three-fourths of 3-A
7. 5-A × 3-A
9. Age of Bob's cousin Eric
10. Half of 12-A
12. 54 × 13-D
14. Days Bob missed school last year

DOWN

1. 64 divided into sum of 4-D and 8-D
2. 12-A subtracted from sum of 8-D and 3-A
3. 1 more than product of 9-D and 10-A
4. Age of father of Bob's Great-Aunt Martha
8. Days Bob has had his paper route
9. Age of Bob's brother Jon
11. 4 × 9-D
13. Prebreakfast snacks Bob had last week

Fill in the clues from the previous page on the puzzle grid below.

3.

<table>
<tr><td>1</td><td>2</td><td>■</td><td>3</td><td>4</td></tr>
<tr><td>5</td><td></td><td>■</td><td>6</td><td></td></tr>
<tr><td>■</td><td>7</td><td>8</td><td></td><td>■</td></tr>
<tr><td>9</td><td>■</td><td>10</td><td></td><td>11</td></tr>
<tr><td>12</td><td>13</td><td></td><td>■</td><td>14</td></tr>
</table>

In the clues, "A" means across and "D" means down. For example, "4-D" would refer to clue number 4 DOWN.

Each square takes a single digit from 0 through 9. No answer begins with 0.

4. Angelo is spending his vacation with his family at a beach house. Angelo said that women don't know anything about fishing, but his Great-aunt Martha proved him wrong yesterday when she caught more fish than he did.

*Reminder: The **square** of a number is the result of multiplying the number by itself. The **cube** of a number is the result of multiplying the number by its square. The **fourth power** of a number is the result of multiplying the number by its cube. And so on.*

ACROSS

1. 5-A ÷ sum of 3-A and 11-A
3. Age of Great-aunt Martha
5. 18-A – 3-A
7. Square of 10-A
9. 3 × 6-D
10. Times Great-aunt Martha played tennis last week
11. Times Great-aunt Martha went surf swimming last week
12. Age of Lucas, Angelo's grandfather
14. Friends who have visited at the beach house so far
16. Rocks Angelo found for Lucas to polish for jewelry
18. Sixth power of 11-A
20. Days Angelo has been at the beach house so far

DOWN

1. Angelo's age
2. Half the sum of 11-D and 1-A
3. Difference between 10-A and 19-D
4. Square of 19-D
6. Fish Great-aunt Martha caught yesterday
8. Cube of 10-A
11. Half the sum of 5-A and 11-A
12. Digits of 7-A reversed
13. 14-A × 15-D
15. Fish Angelo caught yesterday
17. 3-D × 15-D
19. Airplanes Angelo heard overhead while fishing yesterday

Fill in the clues from the previous page on the puzzle grid below.

4.

1	2		3	4	
5		6		7	8
	9				10
11			12	13	
14	15		16		17
18		19		20	

In the clues, "A" means across and "D" means down. For example, "4-D" would refer to clue number 4 DOWN.

Each square takes a single digit from 0 through 9. No answer begins with 0.

5. Casey loves baseball. He has taught his dog, Homer, to retrieve the ball for him when he fires it at a target in the back yard to try to perfect his throwing speed and accuracy. He'd like to take Homer to practices and games, but the manager won't let him.

Reminder: A batting average is a decimal fraction to three places. For example, when we say a batting average is "two thirty-four," the average is really .234 and would have to be multiplied by 1,000 to make it a whole number, 234.

ACROSS

1. Times Casey's shoelaces came undone during games this season
3. Baseball cards in Casey's collection
6. 5 less than 2-D
7. 8-A + 3-D + 20-D
8. Casey's batting average × 1,000
9. 13-D minus 1-A
10. Half of 6-D
11. Casey's uniform number
13. 999 less than 280 × 11-A
15. 27 × 11-A
17. Hours Casey spent last year practicing or playing baseball
19. 11-A subtracted from 10-A
20. Half of 1-D

DOWN

1. Home runs Casey would like to hit before he retires
2. One-third of 8-A
3. 3 less than number formed by first two digits of 12-D
4. Digits of 17-A in a different order
5. Casey's age × 9
6. 8 × result of dividing 1-A by 3
11. Casey's age 21 years from now
12. 18-D × 405
13. 6 less than 2-D
14. Twice the sum of 1-A and 18-D
15. Age of mother of Casey's Great-aunt Martha
16. Digits of 18-D reversed
18. Digits of 16-D reversed
20. Homer's age

5.

	1	2		3	4	5
6				7		
8					9	
10			11	12		
		13			14	
15	16			17		18
19			20			

In the clues, "A" means across and "D" means down. For example, "4-D" would refer to clue number 4 DOWN.

Each square takes a single digit from 0 through 9. No answer begins with 0.

6. Olivia worked several hours on one of her 1,500-piece rectangular jigsaw puzzles before suspecting that something was definitely fishy. Upon investigating, she realized that Luis, her little brother, had exchanged over 25 pieces of it with pieces of two other puzzles of similar coloring. She was angry and said he was too used to getting away with things just because he was the youngest in the family, but he just laughed and said that's what she got for bragging so much about being such a brain at working jigsaw puzzles.

Reminder: The number of pieces printed on the box of a jigsaw puzzle is not usually exact. For example, a rectangular puzzle of "500 pieces" might be 19 × 26 (494 pieces) or 22 × 24 (528 pieces) or some other combination whose product is approximately 500.

ACROSS

1. Half of age of Olivia's Great-aunt Martha
3. Olivia's age
5. Age of Olivia's father
6. Pieces in length of puzzle
7. Pieces Luis exchanged × difference between 1-A and 14-A
9. Hours Olivia spent figuring out what was wrong
10. 13-D + 3-A + half of 3-D
12. 3-D ÷ 11
14. Jigsaw puzzles Olivia worked last month

DOWN

1. Pieces in width of puzzle
2. 3-D ÷ 8
3. Pieces in puzzle
4. 6 more than age of Olivia's Uncle Frank, with digits reversed
8. Age of Leon, Great-aunt Martha's husband × age of Pietro, Olivia's cousin
9. Age of Luis
11. 3-D divided by 17
13. Hours it took Olivia to find and re-exchange puzzle pieces after discovering what was wrong

6.

1	2		3	4
5			6	
	7	8		
9		10		11
12	13			14

In the clues, "A" means across and "D" means down. For example, "4-D" would refer to clue number 4 DOWN.

Each square takes a single digit from 0 through 9. No answer begins with 0.

7. Ismail, Julia, Reynaldo, and Suki are four children who live on Anna's street. Anna doesn't usually see them when she goes to work, but when she's on her way back home they are often outside and she stops to talk to them. The differences between successive ages of the children (starting with the youngest) are 1, 2, and 3 years.

ACROSS

1. Age of Anna's Great-aunt Martha, with digits reversed
3. Reynaldo's age
4. Ismail's age
5. Julia's age

DOWN

1. Anna's age
2. Suki's age
4. 3-A $\times$ half of 1-A
5. Age of Selena, Anna's youngest niece

1	2	■
3	■	4
■	5	

> *In the clues, "A" means across and "D" means down. For example, "4-D" would refer to clue number 4 DOWN.*
>
> *Each square takes a single digit from 0 through 9. No answer begins with 0.*

8. Iris, Jonquil, Lilac, and Marigold each bought a book at a garage sale. One of the books cost 57¢, and the total cost of all four was $1.96. Marigold's book cost the most. From the least paid to the most paid, the prices were equally spaced. (For example, if the lowest two prices were 1¢ and 4¢, then the other two prices were 7¢ and 10¢.)

ACROSS

1. Cousins Jonquil has
2. Cost of Jonquil's book
4. Cost of Iris's book
6. Garage sales Iris went to last month
7. 2-D × 11-D × age of Lilac's Great-aunt Martha
9. Quotient of sum of digits of 3-D and difference between digits of 2-A
10. Cost of Marigold's book

DOWN

1. 11 × 9-A × sum of 2-A, 10-A, and 11-D
2. Brothers Iris has
3. Cost of Lilac's book
5. 1/10 of sum of 3-D and 10-A
8. 11-D × sum of digits of 10-A, 2-A, and 3-D
11. Books Great-aunt Martha bought at this garage sale

1		2	3
4	5		6
7		8	
9		10	11

In the clues, "A" means across and "D" means down. For example, "4-D" would refer to clue number 4 DOWN.

Each square takes a single digit from 0 through 9. No answer begins with 0.

9. Alton, Beaufort, and Carlton are three teenagers who intend to study law. Alton has a tentative acceptance at UCLA, Beaufort at Wayne State University, and Carlton at Harvard University.

Alton is 2 years younger than Beaufort and 1 year older than Carlton. They all work at a restaurant managed by Dalton, the son of Carlton's Great-aunt Martha, who predicts that all three of the young men will have successful careers.

ACROSS

1. Age of Alton's cat, Tinkerbelle
3. Carlton's age
5. Diameter (inches) of trunk of maple tree in Dalton's back yard
6. Alton's age
7. 9-A × 14-A × 13-D
9. Times Great-aunt Martha won at bridge last week
10. Hamburgers restaurant sold for lunch yesterday
12. 10-A + 14-A − 8-D
14. Age of Tinkerbelle's daughter Tinklebelle

DOWN

1. Age of Carlton's brother Elmer, who is 8 years younger than Beaufort
2. 12 divided into product of 1-A, 5-A, and sum of 3-A and 13-D
3. 11-D × 61
4. Age of Great-aunt Martha
8. Sum of ages of all people in this puzzle except Dalton
9. Sum of ages of Alton, Beaufort, and Carlton
11. Beaufort's age
13. Times Great-aunt Martha played bridge last week

Fill in the clues from the previous page on the puzzle grid below.

9.

1	2		3	4
5			6	
	7	8		
9		10		11
12	13			14

In the clues, "A" means across and "D" means down. For example, "4-D" would refer to clue number 4 DOWN.

Each square takes a single digit from 0 through 9. No answer begins with 0.

10. Adam, Benjamin, Cora, and Diane all had different test scores in Mrs. Graham's math class yesterday. All scores were more than 50 and less than 100. The difference between the lowest two scores was 4. The difference between the highest and lowest scores was 24. The difference between the middle two scores was 8. Benjamin did not have the lowest score. Adam's score was higher than Cora's but lower than Diane's.

ACROSS

1. Cora's score
3. 27 × result of dividing 10 less than 1-D into 15-D
7. 18-A divided by 36
9. 1-D + 21-A + 19-D
10. 13 × sum of 13-A, 24-D, and 25-A
12. Benjamin's score
13. Tacos Adam's Great-aunt Martha ate yesterday
14. $\frac{3}{2}$ × one more than 8-D
16. Diane's score
18. Product of lowest two scores
21. Age of Jacob, husband of Great-aunt Martha
22. 5,000 less than product of highest two scores
23. Product of 5, half of 20-D, and sum of 13-A and 24-D
25. Cora's score divided by 12

DOWN

1. Age of Great-aunt Martha
2. 24-D × half the result of subtracting sum of 13-A and 25-A from sum of 21-A and 24-D
3. Years Carmen, brother of Diane, has had a paper route
4. Average golf score (18 holes) of Lachlan, brother-in-law of Great-aunt Martha
5. Total of all four test scores
6. 4 × sum of 4-D and age of Lachlan's son Kirby
8. Age of Lachlan's son Charles subtracted from half of 6-D
11. 18-A + 22-A + 5,000
13. 7 × 1-D
15. Number of different test problems Mrs. Graham has used in the last two years
17. 8 × 4-D
19. Adam's score
20. 3 × 4-D
24. Tests Edward, nephew of Mrs. Graham, helped Mrs. Graham correct last month

10.

1	2		3	4	5	6
7		8		9		
	10		11		12	
13		14		15		
16	17		18		19	20
21			22			
	23	24				25

In the clues, "A" means across and "D" means down. For example, "4-D" would refer to clue number 4 DOWN.

Each square takes a single digit from 0 through 9. No answer begins with 0.

11. Grayson's supermarket is only a quarter of a mile from Daniel's home. He doesn't mind being sent there to do some shopping, for he enjoys discovering new products on display and talking to other people who regularly shop there.

The last time Daniel went to Grayson's he bought seven items: bread, gravy mix, hamburger, hand soap, onions, spaghetti, and tomato sauce. The prices he paid for these items were (in cents) 33, 49, 54, 63, 87, 98, and 148.

ACROSS

1. See 2-D
3. Cost of hand soap
5. Cost of hamburger
7. Cost of spaghetti
9. See 2-D
10. Age of Daniel's sister Pamela
11. Times Chaser, the family's dog, went outside and then wanted back inside yesterday afternoon
12. Cost of gravy mix
14. Cost of tomato sauce
16. 17-D + 20-A + 22
18. 9-A – 1-A + 3-D + 19-D + three times age of Daniel's Great-aunt Martha
20. See 16-A

DOWN

1. 3-D + 10-A + 11-A + 19-D
2. 1-A + 9-A + 241
3. Times Daniel went to Grayson's last week
4. Cost of bread
6. 7 × sum of 3-D and 10-A
8. Cost of onions
11. Total cost of all seven items
12. 5-A minus 8-D
13. Sum of 4 × 1-A and 3 × 9-A
15. One-fourth of 5-A
17. See 16-A
19. Times Daniel mowed front lawn last month

Fill in the clues from the previous page on the puzzle grid below.

11.

1	2		3	4	
5		6		7	8
	9				10
11			12	13	
14	15		16		17
18		19		20	

In the clues, "A" means across and "D" means down. For example, "4-D" would refer to clue number 4 DOWN.

Each square takes a single digit from 0 through 9. No answer begins with 0.

12. Dal-Mei's mother gave an outdoor birthday party for Dal-Mei yesterday afternoon. Between playing games, the guests enjoyed helping themselves to the snacks provided. The party ended after the guests were served sandwiches, lemonade, and Dal-Mei's birthday cake, which was decorated with good luck symbols and held a candle for each year of Dal-Mei's age, plus one extra candle for good luck.

ACROSS

1. One less than the quotient of 7-A and 7
3. Guests at party
5. 5 divided into sum of 8-D and 4
6. Hours Dal-Mei's mother spent preparing for party
7. Last three digits of 3-D, in reverse order
9. Age of Dal-Mei's Great-aunt Martha 60 years ago
10. Number of balloons at party
12. 600 less than 2-D
14. Dal-Mei's age

DOWN

1. Age of Dal-Mei's brother, Dal-Tan
2. Polka dots in Dal-Mei's party dress
3. 10-A × 3-A
4. Candles on birthday cake
8. Peanuts eaten at party
9. 5-A + 11-D
11. Dal-Mei's age in ten years
13. 3-A divided by 7

Fill in the clues from the previous page on the puzzle grid below.

12.

1	2		3	4
5			6	
	7	8		
9		10		11
12	13			14

In the clues, "A" means across and "D" means down. For example, "4-D" would refer to clue number 4 DOWN.

Each square takes a single digit from 0 through 9. No answer begins with 0.

13. Timothy's Great-aunt Martha, her husband, Amos, and her brother-in-law, Casper, attended a family gathering at Timothy's home last Sunday afternoon. They all had great fun playing boccie in the back yard before sitting down to the excellent dinner Timothy's parents had prepared.

ACROSS

1. Times Great-aunt Martha went ice-skating last winter
2. See 3-D
4. Sum of digits of 2-A and 3-D
6. Other guests at dinner
7. Sum of ages of Amos and Casper and square of Timothy's age
9. First cousins of Timothy
10. Product of last digits of 5-D and 8-D, with product digits in reverse order

DOWN

1. Product of the four digits in 2-A and 3-D, but with resulting answer digits in reverse order
2. Grandchildren of Casper's
3. 2-A + 19
5. 2 × 8-D
8. Boccie points Timothy's side scored
11. Boccie points Timothy scored for his side

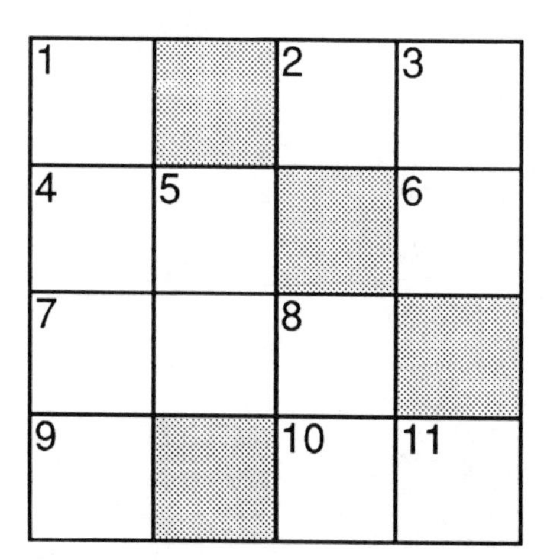

In the clues, "A" means across and "D" means down. For example, "4-D" would refer to clue number 4 DOWN.

Each square takes a single digit from 0 through 9. No answer begins with 0.

14. Ariadne's Great-aunt Martha, who used to be an accountant and is now an electronics engineer, has always enjoyed solving puzzles and working with numbers. She made up this puzzle for Ariadne to solve, promising to play tennis with Ariadne next Saturday if Ariadne could solve the puzzle within twenty minutes.

ACROSS

1. Maple trees in front yard of Sean, Great-aunt Martha's next-door neighbor
2. 2 × 3-D
4. $\frac{3}{2}$ × 3-D
6. See 8-D
7. Square feet in Sean's garden
9. Lilac bushes in Sean's back yard
10. 15 more than age of Great-aunt Martha

DOWN

1. 2-A × 3-D
2. Age of Sean's car
3. See 2-A
5. 2-A + 4-A
8. 10 more than 3-D
11. 5-D ÷ number formed by last two digits of 7-A

1		2	3
4	5		6
7		8	
9		10	11

In the clues, "A" means across and "D" means down. For example, "4-D" would refer to clue number 4 DOWN.

Each square takes a single digit from 0 through 9. No answer begins with 0.

15. John's great-aunt Martha and four of her friends are in a bowling league. The average of their ages is 68.4, and the differences between successive ages (starting with the youngest) are 3, 3, 4, and 3 years. Mr. Young is older than Mr. Xavier but younger than Great-aunt Martha.

The average of their five bowling scores during one game yesterday was 167.2, and successive differences between the scores (starting with the lowest) were 18, 32, 36, and 11. The scores were in opposite order of the ages. That is, the highest score was bowled by the youngest person, and so on, down to the lowest score bowled by the oldest person. Mrs. Wilson's score was better than Mr. Young's score but not as high as the score of Mrs. Santos.

Reminder: The highest bowling score obtainable is 300.

ACROSS

1. Age of Mrs. Zantos
3. 19-D × half of 9-A
7. 34 + 8-D + 10-A
9. Score of Mrs. Zantos
10. See 7-A
12. Age of Mr. Young
13. Number of pizzas the five bowlers split last night
14. Total of the five scores
16. Age of Mr. Xavier
18. 13-A + product of 102 and 21-A
21. Age of Great-aunt Martha
22. 2,000 + quotient of 11-D and 7
23. 10 less than product of 7 and 10-A
25. Number of spares Mr. Xavier made

DOWN

1. Age of Mrs. Wilson
2. Score of Mr. Xavier
3. Number of spares Great-aunt Martha missed
4. John's age
5. Score of Great-aunt Martha
6. Total of ages of the five senior citizens
8. See 7-A
11. 2-D × 5-D
13. Score of Mrs. Wilson
15. 9-A × half of 1-A
17. 9-A + 2-D + 20-D
19. Mr. Young's age when he first started bowling
20. Score of Mr. Young
24. Number of strikes Mrs. Wilson had

15.

1	2		3	4	5	6
7		8		9		
	10		11		12	
13		14		15		
16	17		18		19	20
21			22			
	23	24				25

In the clues, "A" means across and "D" means down. For example, "4-D" would refer to clue number 4 DOWN.

Each square takes a single digit from 0 through 9. No answer begins with 0.

16. Alan, Barbara, Jeremy, Kemil, and Mandy all live on Pine Street and are different ages. The youngest person is older than three, and the oldest person is younger than twenty-one. Jeremy is older than Alan, who lives between Kemil and Mandy. Jeremy lives between Alan and Barbara, while Mandy lives between Kemil and Jeremy.

Jeremy's Great-aunt Martha, who lives on Maple Street, took Jeremy golfing yesterday.

ACROSS

1. Mandy's age
3. Mandy's age × Barbara's age
5. 72 × 11-A
7. Sum of all five ages
9. 14 × this number = 11-A × 6-D
10. Barbara's age
11. Years Kemil has lived on Pine Street
12. Golf score of Great-aunt Martha (18 holes)
14. Sum of ages of Barbara, Jeremy, and Kemil
16. 5-A minus 9-A
18. Jeremy's age × Barbara's age
20. Half of Jeremy's golf score yesterday (18 holes)

DOWN

1. Sum of Barbara's and Kemil's ages
2. Jeremy's age × sum of ages of Alan and Mandy
3. Kemil's age
4. Jeremy's age
6. See 9-A
8. Sum of all ages except Barbara's
11. Product of Jeremy's usual 9-hole golf score and sum of Kemil's and Alan's ages
12. 6-D minus 3-D
13. Mandy's age × Jeremy's age
15. Alan's age
17. Twice the difference between 12-A and 8-D
19. First cousins Mandy has

Fill in the clues from the previous page on the puzzle grid below.

16.

1	2		3	4	
5		6		7	8
	9				10
11			12	13	
14	15		16		17
18		19		20	

Detailed Solutions

[Continued from page iv.]

43, 35 and 54, 46 and 65, 57 and 76, 68 and 87, 79 and 98. Because 1-D has four digits and because none of these digits can be 0 (no answer can start with 0), 1-D eliminates the first five possibilities. So 2-A, 3-D, and 1-D are, respectively, either 68, 87 and 8862; or 79, 98, and 6354. 4-A eliminates the first of these, forcing answers to 2-A, 3-D, 1-D, 4-A, 1-A, 2-D, 6-A, and 9-A.

5-D and 8-D must be 30 and 15, 32 and 16, 34 and 17, 36 and 18, or 38 and 19. However, 10-A eliminates all but 36 and 18, thus forcing answers for 5-D, 8-D, 10-A, 7-A, and 11-D.

14. 3-D is an even number less than 50 (see 2-A and 4-A). So 3-D is in the range 12–48. Combining 1-D and 2-A implies that 1-D is twice the square of 3-D. 1-D must be at least 1,112, so the square of 3-D must be at least 556. This puts 3-D in the range 24–48. Since the first digit of 3-D is the second digit of 2-A, the only possibilities for 2-A and 3-D are 52 and 26, or 84 and 42. Then 1-D is either 1,352 or 3,528. But answers of 84, 42, and 3,528 don't work for 4-A. Thus answers of 52, 26, and 1,352 are forced for 2-A, 3-D, and 1-D, in turn forcing 1-A, 2-D, 6-A, 9-A, 4-A, 8-D, 5-D, 7-A, 11-D, and 10-A.

15. The average of the five ages was 68.4, so their total was 68.4 × 5, or 342 (6-D).

    If the students do not know algebra, we can reason as follows: Given successive age differences of 3, 3, 4, and 3 years, the oldest was 13 years older than the youngest, so we can subtract 13 from 342 (=329) and have the sum of twice the age of the youngest and the ages of the next three oldest. Similarly, we subtract 10 and then 6 and then 3, ending up with five times the age of the youngest = 310. It follows that the ages are 62, 65, 68, 72, and 75.

    By analogous reasoning, 14-A is 836, and the scores are (order corresponding to ages) 214, 203, 167, 135, and 117.

    Since Mr. Young's score was less than those of Mrs. Wilson and Mrs. Zantos, Mr. Young is older than both of these people. He is older than Mr. Xavier and younger than Great-aunt Martha, so Great-aunt Martha is the oldest (and had the lowest score). Mr. Young is the second oldest (and had the second lowest score).

    This forces answers to 6-D, 14-A, 21-A, 5-D, 12-A, 20-D, 25-A, 13-D (only one score ending in 7 is left), 13-A, 1-D (age placed by score), 1-A (cannot be 65 since 2-D is a bowling score), 9-A, 16-A, 2-D, 15-D, 11-D, 22-A, 18-A, 19-D, 3-A, 3-D, 4-D, and 17-D.

    The last digit of 8-D + 10-A is 0. When 34 is added to the sum, the last digit will be 4. Therefore, 7-A is 804, and 8-D + 10-A = 804 – 34 = 770. With a middle digit of 0, the sum is 302 + 408 = 710 (60 short of 770) so the middle digit is 3, forcing answers to 23-A and 24-D.

16. Two ages are under 10 (10-A, 3-D), and the other three are 10 or more (1-A, 4-D, 15-D). For 7-A, the least sum possible is 42 (4 + 5 + 10 + 11 + 12), while the greatest sum possible is 74 (8 + 9 + 18 + 19 + 20). But this sum minus 10-A = 8-D, and a process of elimination will show that 7-A, 8-D, and 10-A are 54, 47, and 7. (To shorten the process, consider that the second digit of 7-A has to be either the same as, or one less than, the first digit. This eliminates all possibilities for 7-A except 43, 44, 54, 55, 65, and 66. Then try 10-A ranges from 4–9.)

    Jeremy's age (4-D) has to be 15. For 3-A, the only 2-digit numbers ending in 1 and having a factor of 7 (from 10-A) are 21 (3 × 7) and 91 (13 × 7). Mandy can't be 3, so she is 13, and so answers are forced for 1-A, 3-A, 3-D, 15-D (54 – sum of other four ages), 1-D (3-D + 10-A), 14-A (1-D + 4-D), 18-A (4-D × 10-A), 19-D, 2-D (4-D × sum of 15-D and 1-A), and 13-D (1-A × 4-D).

    5-A is in the 640s and has 72 as a factor, so 5-A = 72 × 9 = 648, forcing answers to 11-A and 11-D.

    11-A is 9, so 9 is a factor of 9-A. The only number in the 50s having a factor of 9 is 54, so answers are forced for 9-A, 6-D, 16-A, 12-D, 12-A, 17-D, and 20-A.

    The positions of the houses do not affect the solution, but for those who are interested, the order is: Kemil, Alan, Mandy, Jeremy, Barbara; or vice versa.